Ambulatory Surgery Center Safety Guidebook

Ambulatory Surgery Center Safety Guidebook

Managing Code Requirements for Fire and Life Safety

Dale Lyman

Butterworth-Heinemann
An imprint of Elsevier

Butterworth-Heinemann is an imprint of Elsevier
The Boulevard, Langford Lane, Kidlington, Oxford OX5 1GB, United Kingdom
50 Hampshire Street, 5th Floor, Cambridge, MA 02139, United States

Notices
Knowledge and best practice in this field are constantly changing. As new research and
experience broaden our understanding, changes in research methods, professional practices, or
medical treatment may become necessary.

Practitioners and researchers must always rely on their own experience and knowledge in
evaluating and using any information, methods, compounds, or experiments described herein.
In using such information or methods they should be mindful of their own safety and the safety
of others, including parties for whom they have a professional responsibility.

To the fullest extent of the law, neither the Publisher nor the authors, contributors, or editors,
assume any liability for any injury and/or damage to persons or property as a matter of products
liability, negligence or otherwise, or from any use or operation of any methods, products,
instructions, or ideas contained in the material herein.

Library of Congress Cataloging-in-Publication Data
A catalog record for this book is available from the Library of Congress

British Library Cataloguing-in-Publication Data
A catalogue record for this book is available from the British Library

ISBN: 978-0-12-849889-7

For information on all Butterworth-Heinemann publications visit our
website at https://www.elsevier.com/books-and-journals

Working together
to grow libraries in
developing countries

www.elsevier.com • www.bookaid.org

Publisher: Candice Janco
Acquisition Editor: Candice Janco
Editorial Project Manager: Hilary Carr
Production Project Manager: Punithavathy Govindaradjane

Typeset by SPi Global, India

CONTENTS

CHAPTER *1*

Fire and Life Safety in Ambulatory Surgery Centers

INTRODUCTION

There are over 5260 ambulatory surgery centers (ASCs) in the United States, and 5723 hospitals, and these numbers are increasing each year. Technological and cultural trends in health care have shifted the majority of surgery to the outpatient environment, and an aging population is also driving the growth in demand for surgical procedures in general. Similar to other health care organizations, ASCs are highly regulated by federal and state entities. The fire and life safety in ASCs is evaluated by independent observers, inspectors, and surveyors for state licensure, Medicare certification, and voluntary accreditation; therefore code compliance is essential for a facility's existence and survival.

VULNERABILITY OF PATIENTS AND RISK

Fire and life safety codes place ASCs under intensive scrutiny because of the vulnerability of the patient population occupying these facilities. The procedures performed are often comparable to those performed in surgical suites in hospitals; the difference is that ASC patients are discharged after a short recovery period following their procedure, with no overnight stay at the facility. Because the patients are anesthetized and their physical mobility is impaired, they would be unable to escape from a dangerous situation such as a fire. They may even be on life support equipment and being monitored by critical care personnel. During a given period of time, these patients are totally dependent on staff members for physical assistance to evacuate them and to reach safety in an emergency situation.

Dangerous Environments

Fires in health care environments are not uncommon. Between 2006 and 2010, an average of 6240 fires occurred in medical facilities each year, resulting in six deaths, 171 injuries, and $52.1 million in direct property

Ambulatory Surgery Center Safety Guidebook. https://doi.org/10.1016/B978-0-12-849889-7.00001-7

damage annually in the United States.[1] Codes and regulations are very stringent in ASCs for two reasons. The first, as mentioned above, is because many patients may not have the physical ability to escape a fire. The second is due to the many fire and life safety hazards found within health care environments. These two conditions combine to create a very high level of risk. This is why the strict set of codes has been written, to mitigate those risks. One of those hazards is the medical gasses found in an ASC. Oxygen, which is not flammable as a single element but acts as an oxidizer and a catalyst to the combustion process, can contribute to extremely dangerous, fast-moving fires. In other words, oxygen will greatly increase the intensity of the combustion process (fire). Even inert gasses, such as nitrogen, commonly found in health care environments could create a dangerous scenario if they were to leak within an enclosed space by displacing and reducing the oxygen level, as this would cause a risk of asphyxiation.

Surgical procedure sites and fields are ripe for an unwanted, unexpected fire. Because all three basic elements necessary for combustion are present—fuel (alcohol preps, paper dressing, drapes), oxygen (concentrated medical oxygen), and an ignition source (often lasers)—a flash fire can occur unexpectedly during a critical procedure. A power failure during surgery could have catastrophic consequences if a facility is not equipped with a reliable backup power supply to allow the procedure to be concluded safely. Even products that are crucial to control infections (such as alcohol-based hand rub), which are a necessary part of patient care to ensure infection control, are classified as Class 1 flammable liquids. This is because of their volatile properties, which greatly contribute to the potential for a fire. This overall combination of extremely volatile materials and hazards found in an ASC, coupled with the surgical patients, who are among the most vulnerable to occupy any building, results in one of the most stringent set of building and fire codes found in any structure.

REGULATION OF HEALTHCARE FACILITIES

All ASCs serving Medicare beneficiaries must be certified by the Medicare program, and are inspected or "surveyed" for compliance with fire and life safety codes. In order to ensure that a facility meets the requirements, the

---

[1] http://www.nfpa.org/research/reports-and-statistics/fires-by-property-type/health-care-facilities/fires-in-health-care-facilities, Marty Ahrens, November 2012, NFPA Topical Report, "Fires in Health Care Facilities."

ASCs are inspected or "surveyed" for state licensure or Medicare certification. Alternatively, an ASC may opt for a "deemed status" survey, to be provided by a third party accrediting agency that has entered into a contractual agreement with the federal government to ensure that facilities are in compliance with federal regulations. Becoming accredited as a Medicare certified facility and thus being allowed to participate in its billing program can have tremendous financial consequences for an ASC. The Centers for Medicaid and Medicare Services (CMS) has determined that facilities must adhere to their set of rules for patient safety; these rules are referred to as "Conditions for Care."

CONDITIONS FOR CARE

The Conditions for Care pertaining to the physical environment of an ASC state that the facility must be a "distinct entity." This requirement includes rules for separate waiting rooms for an ASC from other medical or clinical patients, and the availability of restroom facilities for waiting room occupants. One condition for coverage specifies that operating rooms and procedure rooms must be designed to the level of current standards as well as meeting specific requirements for temperature, humidity, and airflow. Another condition addresses patient privacy. One of the most detailed conditions establishes specific criteria for the use of alcohol-based hand rub dispensers in ASC facilities. These requirements eventually became part of fire and life safety codes; they resulted from several unfortunate incidents where alcohol-based hand rubs were found either to be contributing factors to the cause of actual fire incidents or to contribute significantly to the spread of a fire because of their highly flammable properties. Chapter 11 of this guidebook further examines the requirements associated with alcohol-based hand rubs.

Federal Fire and Life Safety Code Requirements

This guidebook focuses on Medicare condition for coverage in Section 416.44(b) Environment. This section specifies that the ASC (Ambulatory Surgery Center) must meet the provisions applicable to Ambulatory Health Care Centers found in the year 2000 edition of the Life Safety Code, published by the National Fire Protection Association (NFPA), which was adopted by federal law in March 2003. The Code covers a wide range of requirements related to reducing the spread of fire within buildings and ensuring a means of egressing a building during an emergency event. Because NFPA 101 is a general fire and life safety code, it refers to subsequent special topic codes to address the unique and specialized equipment and systems found within an ASC. These subsequent codes include:

- NFPA 99 Health Care Facilities Code, which addresses specific risks found in health care facilities;
- NFPA 110 Emergency and Standby Power Systems and NFPA 111 Stored electrical Energy Emergency and Standby Power Systems, which address emergency power requirements;
- NFPA 13 Installation of Sprinkler Systems, which provides information regarding a facility's fire sprinklers; and
- NFPA 72 National Fire Alarm and Signaling Code, which addresses fire alarm system requirements.

There are portions and sections of all of these codes that an ASC must adhere to in order to remain in compliance, as well as the ongoing inspection, testing, and maintenance of the building and required specialized systems. Unfortunately, many ASC facility personnel are not familiar with all of these ongoing "operational" requirements, and this can lead to noncompliance issues during licensure or accreditation surveys and inspections.

TOTAL CONCEPT CODE REQUIREMENTS

The NFPA 101 Life Safety Code is considered a "total concept" code for health care facilities in that it is based on the premise that these facilities be designed, constructed, maintained, and operated to minimize the possibility of an emergency event that would require the occupants to evacuate. Because of the occupants' physical condition, their safety cannot solely rely on evacuation, as may be the case in other types of buildings, and therefore the code must employ a tiered or layered system of strategies to achieve maximum life safety. These strategies include:

(1) building design, construction, and compartmentalization;
(2) active fire protection systems that include fire detection, alarm, and extinguishment; and
(3) fire prevention planning and procedures, training of staff and fire drills, planning, and training of staff to relocate occupants to areas of refuge.

In other words, there are redundant layers of protection to ensure the safety of the occupants. For example, if a fire were to occur in an ASC, the first layer of defense (or first component of the total concept) providing protection for the occupants would be the construction features of the building itself and its built-in systems. The fire would be contained and its spread would be limited by fire walls and fire suppression systems. The second layer/component is that those required features or systems are maintained properly to ensure that they will perform as designed. The third layer/component for occupants' protection is that the facility's staff are trained properly and the facility has policies and procedures for its staff that will ensure the safety of both staff and patients.

Structural and Building Requirements

When a facility is constructed, professionals such as architects, engineers, and designers who are all familiar with fire and life safety code requirements are retained to complete the project. Therefore, when a facility is surveyed for fire and life safety code compliance, there are often no areas of noncompliance with the physical/building requirements, because these professionals are familiar with the required "built-in" design and fire protection components that are required by the code. So this accounts for components 1 (building design) and 2 (active fire protection features) of the total concept being in compliance. However, once the facility is opened and turned over to the medical and health care professionals, there may be a lapse in code requirements from an operational perspective, causing a facility to be in noncompliance. This may be because the facility's staff are unaware of the third component of the "total concept," the operational requirements, which includes fire prevention planning/training and policies, and inspection, testing, and maintenance of required systems.

Administrative and Operational Requirements

This third component of the "total concept" requires a facility's staff to have all the required administrative and operational policies in place, train staff on the policies, and keep accurate, up-to-date records of these activities. Certain components of the overall building, or physical environment, must also be inspected, tested, and maintained at code-required intervals. Complete and accurate documentation must be retained in order for the facility to be fully compliant with the fire and life safety codes. All of this falls under the operational requirements.

OBJECTIVE OF THIS BOOK

The objective of this guidebook is to provide a basic understanding of the overall fire and life safety code requirements for staff members of ambulatory surgery centers—specifically, those staff members who are responsible to ensure their facility meets all necessary requirements for licensing and CMS and/or CMS deemed status surveys. In order to maintain compliance, it is crucial that staff possess a basic understanding of their facility's physical environment components, such as fire separation walls, fire alarm systems, and fire sprinklers, and the extremely important tasks involved with inspecting, testing, and maintaining these systems. Staff must also have a solid foundational knowledge of the administrative and operational requirements that span from policies and procedures that specify actions and staff member roles during an emergency event, to training and drilling staff in the facility's emergency policies.

This guidebook will provide example templates to ensure that all required operational and documentation requirements are completed. The fire and life safety codes require regular inspection, testing, and maintenance at specified intervals. In order to maintain code compliance, these records must be kept on site and be readily available for review by inspectors and surveyors. The sample templates in this guidebook include the required intervals, which may be daily, weekly, monthly, or annually. The documentation templates are presented in a systematic framework to ensure that all required components are accounted for and are completed in the required time frames and ensures that all three components of the total concept are completed and that code compliance is achieved.

What This Book Is Not

It must be emphasized that this guidebook is not a substitute for the Life Safety Code or any locally adopted fire or life safety codes; a facility's staff should refer to specific code books if they have questions or are in need of guidance for exact requirements. This guidebook should not be considered a code "handbook," as it is the intent not to provide code interpretations, or detailed explanations, but rather to provide guidance for health care and administrative professionals who do not have a code interpretation background. The information and templates provided in this guidebook will provide an understanding and tools to help prepare for inspections and surveys. This guidebook assumes that the reader's facility is open, operating, and meets all the physical, built environment requirements of applicable building, fire, and safety codes of the applicable jurisdictions. This guidebook should not be considered as a design guide for those preparing to construct a facility. Always consult with the "Authority Having Jurisdiction" at the federal, state, and local levels for specific fire and life safety code requirements for your specific facility, to ensure compliance.

CHAPTER 2

Life Safety Code and the "Total Concept"

TOTAL CONCEPT

There have been a number of tragic fire incidents in the United States and throughout the world, in which many innocent people have lost their lives, which have led to the creation and evolution of the fire and life safety codes that exist today. In the United States there are two main code development organizations, which utilize a democratic process of code development and writing in which members of the organizations actively submit proposals and/or changes. The overall memberships then have an opportunity to comment and ultimately vote on the proposals. The final codes are published and government entities at the local, state, and federal levels may choose to adopt the published code as law. In the United States, the two main code-making organizations are the International Code Council (ICC) and the National Fire Protection Association (NFPA). Therefore, an ASC may be subject to codes adopted at each of these levels of government.

The code selected by the federal-level CMS is the Life Safety Code, because it is a very broad, all-encompassing set of rules and requirements. Certain aspects of this Code are found in building codes and also in fire codes. The Life Safety Code has been referred to as a total safety concept code.

It goes beyond the physical aspect of how a building is physical built (materials, design) and includes requirements for inspection, testing, and maintenance of components. Additionally, it includes requirements for preparing the occupants to respond appropriately to an emergency event. In order for an ASC to be compliant with fire and life safety code requirements, it must meet not only the physical environment requirements, but also a set of administrative and documentation requirements. All of these components and requirements work together.

Ambulatory Surgery Center Safety Guidebook. https://doi.org/10.1016/B978-0-12-849889-7.00002-9

PHYSICAL ENVIRONMENT REQUIREMENTS/COMPONENTS

The physical environment requirements of fire and life safety codes include how the facility is built, include the actual materials used for construction, how it is designed, and the specific equipment and systems installed in the building in order to maximize the protection and life safety of the people who will occupy the building.

Overall Structure

Fire and life safety codes contain specific requirements regarding how a building can be constructed. In the case of an ASC, the codes specify what type of construction materials can be used, how they are assembled, and even what type of other tenants may share a building that an ASC occupies. The intent of these requirements is to limit fire ignition and, if a fire does occur, to limit its spread and effect on the overall structural integrity of the building. The codes also contain many requirements for compartmentalization, which is a strategy of dividing a building or facility into smaller, sealed compartments through the use of fire walls or fire barriers in order to keep fire and smoke from spreading throughout the interior of a building. This concept will be discussed in depth in Chapter 3. ASCs have specific code requirements pertaining to the overall building in which they are located to include certain compartmentalization requirements and separations achieved with firewalls and barriers.

Egress or Exiting

Providing a "way out" or means of egress is one of the most important components of fire and life safety codes. ASCs have specific requirements for the number of exits, the location and configuration of those exits, and the overall design of those exits depending on the size of an ASC and the floor or story of a building where that facility is located. A number of sub-components of an exit must also meet specific requirements, such as width, signage and identification, and door hardware used to operate latching mechanisms. The code also takes into consideration the safety of an exiting occupant once that person is outside the structure. All codes include requirements for the exiting occupant to safely reach the "public way," which is defined as a street, alley, or another parcel of land. The intent is that the exiting occupant will be able not only to exit, but to move to a safe distance from the burning structure, in order to be completely out of harm's way.

Fire Protection Systems

The historical performance of fire protection systems, such as fire sprinklers, fire alarms, and fire extinguishers, has demonstrated their added value in increasing life safety as part of a building design. Because of this en-hanced level of life safety, fire and life safety codes require these systems to be present in buildings occupied by facilities such as an ASC because of its highly vulnerable population. These systems act as a first line of defense in a fire scenario, and give the occupants more time to evacuate safely and move to an area of refuge. The prompt use of a fire extinguisher on a small, incipient fire can change the outcome of an incident and prevent a tragedy. Fire alarm systems can provide early detection of a fire, even in unoccupied rooms or parts of a building, and then, most importantly, notify occupants throughout the building that evacuation is required and also automatically notify the Fire Department. An ASC may or may not be required to have fire sprinklers installed, according to the size of the facility and the story or floor of the building in which it is located. A fire sprinkler system is a series of nozzle-like devices, referred to as sprinklers; these are connected to piping and arranged so that they will automatically distribute sufficient quantities of water to extinguish a fire. Heat from a fire causes affected sprinklers directly above the fire to open and emit water. Fire and life safety codes include specific requirements for fire protection systems in an ASC.

Emergency Power Systems/Supplies
There are times when patients in an ASC depend on external support and critical care equipment to sustain life. It is therefore crucial that this critical care equipment is designed to ensure the continuity of electrical service to specific areas and pieces of equipment when there is a disruption in the normal source of power. Fire and life safety codes thus include a number of requirements related to emergency power and/or backup power supplies. These backup power supplies and systems are referred to as "essential elec-trical systems." The codes define specific types or levels of equipment based on the types of medical procedures performed and level of anesthesia that is used in a facility. The requirements are often achieved by utilizing a gen-erator; however, an ASC may be able to comply with code requirements by using a battery system, or equipment backed up by battery.

Medical Gasses/Distribution Systems
In order to reduce the risks associated with the use and distribution of med-ical gasses, fire and life safety codes include a number of requirements that ASCs must adhere to. Oxygen, which by itself is not flammable, acts as a catalyst and exponentially increases the spread and intensity of a fire. Inert gasses such as nitrogen are not flammable, but pose a health hazard in that an unwanted release has the potential of displacing the oxygen within a

given area and creating a risk of asphyxiation. There are also hazards and risks associated with piped medical gas systems, such as large numbers of gas cylinders being stored in one concentrated location and medical gas piping passing through walls. To mitigate these risks, the codes include requirements for storage, handling, and distribution systems.

Medical Equipment

The medical equipment contained within an ASC is vital to a facility's operation. Most medical equipment is electronic and requires electricity. The hazard and concern is the potential of exposing a patient to electric shock. If equipment is not installed properly and not maintained to standards defined in fire and life safety codes, it becomes a risk to the very patients it is intended to treat. Once again, depending on the type of procedures performed in a facility, there may be requirements in surgical suites for protecting staff and patients from electrical shock via specific types of safeguarding equipment, such as line isolation monitoring and/or ground fault interrupter circuits. Additionally, there are specific requirements for electrical tools, power cords, and electrical medical equipment to undergo regular testing to ensure electrical integrity.

Alcohol-Based Hand Rub Dispensers

Strict infection control policies and procedures are crucial to the operation of an ASC. Best practices in the area of infection control include readily available alcohol-based hand rub dispensers. However, because of the highly flammable properties of these products, fire and life safe codes contain a number of elements that must be adhered to in order to remain in compliance with the codes. These include restrictions on where they can be mounted, distance to another mounted container, and storage requirements for products. These rules were derivative of incidents that occurred shortly after the influx of alcohol-based hand rubs in health care facilities. Because of the high fire risk of these hand rubs, the codes quickly adopted requirements in order to ensure safety regarding their use.

Interior Finishes

The interior finishes of an ASC are regulated by fire and life safety codes to help control the spread of fire in a facility. A flammable or combustible interior finish in a facility can result in a quickly spreading fire. Review and study of past incidents has shown this to be the case, and the codes attempt to mitigate this hazard through specifying interior finish materials based on their flame spread characteristics. The codes address materials used for curtains, privacy curtains, decorative materials, and furnishings.

OPERATIONAL AND DOCUMENTATION REQUIREMENTS

In addition to the physical requirements that fire and life safety codes require for an ASC, a facility must also comply with a set of administrative and documentation requirements. An ASC needs to have administrative policies in place to ensure that all required systems are inspected, testing, and maintained in accordance with requirements defined in the codes. Additionally, administrative policies and procedures are required to ensure that a facility's staff will react properly to a fire or other emergency event. These include requirements for familiarity with emergency policy and procedures, training, and drills to exercise and "put into motion" these policies and procures. Finally, one of the most important components of these requirements, demonstrating compliance, involves the documentation of all of the above. These administrative and documentation requirements help to complete the overall total safety concept and ensure a comprehensive life safety strategy.

Inspection, Testing, and Maintenance

Fire and life safety codes require that all fire and life safety systems installed within a facility are inspected, tested, and maintained to code standards. The intent is to ensure that the systems will be ready to perform their intended functions should a fire or other emergency occur. Proper maintenance helps to keep these systems in good working order and may be used to address any deficiencies identified through the required inspections and testing. Specialized tools and technical training are typically required to perform maintenance, which is why codes require the maintenance to be performed by qualified professionals. It is important that the ASC recognizes the importance not only of required maintenance being performed but also of keeping documentation of all maintenance readily available, in order to demonstrate code compliance.

Occupant Activities, Preparation, and Training

Ensuring that the staff of an ASC respond appropriately during a fire or other emergency will greatly affect the outcome and the overall safety of staff and patients. It is imperative that facilities have operational policies that address managing a fire as well as managing the occupants. The policies should specify actions taken by staff when a fire is discovered, and then identify the actions to take to isolate patients from the fire and then to isolate or evacuate them from the fire. Codes also require a facility to identify the staff responsible for inspection, testing, and maintenance of systems.

Written Policies and Procedures/Drills

Once policies are established by a facility. These must be put into action through exercises and drills. Fire and life safety codes set out specific requirements for frequency of drills and criteria for documenting the performance of staff during the drills. Documentation is important because it not only demonstrates code compliance but can also identify trends in staff performance. This can help to indicate areas where a plan may need to be modified to meet the realistic abilities of an ASC's staff. ASCs are required to maintain records of exercises and drills to remain in compliance with fire and life safety codes.

Documentation

The importance of on-site documentation cannot be stressed enough. Completed and thorough documentation is required for a facility to be compliant with fire and life safety codes. Documentation for all items relating to code compliance should be kept in a single three-ring binder or documentation scanned into a database, and should be readily accessible to inspectors or surveyors. Being organized can speed up the document review portion of inspections and surveys. Keeping all documentation in one place also reduces the likelihood of it being lost or misplaced, which could result in a code violation or deficiency rating.

Health care settings, and specifically ASCs, are unique environments with unique occupants who are not always capable of self-preservation during a fire or other emergency. All of the required areas of code compliance in the ASC's environment, including physical, administrative, and documentation requirements, complement each other and contribute to the total safety concept. All of the items described above will now be described and explained in more detail to give ASC staff a foundational understanding of these systems and requirements, in order to ensure a safer, code-complaint ASC.

CHAPTER *3*

Building Construction Requirements

BUILDING CONSTRUCTION REQUIREMENTS

Building and fire codes incorporate two basic strategies to protect the occupants of buildings, should they become involved in a fire. One is evacuation, which includes requiring egress (or exiting) components in design and construction, as well as early warning systems and notification, such as fire detection and alarm systems, utilized in order to initiate evacuation. These concepts will be discussed in later chapters. The second strategy is called "defend in place," and recognizes the fact that non-ambulatory occupants, such as a patient in an ambulatory surgery center, are actually subjected to more risk when staff attempt to evacuate them from a structure. Depending on where a fire is located within a building, it may be safer for the patient to remain in their current location, or perhaps to be relocated to another nearby location within the building. Additionally, this strategy recognizes the fact that these patients are unable to evacuate a building on their own, and require assistance and possibly life support equipment.

Compartmentalization Concept and "Defend in Place" Strategy

In order to create an environment or build a structure that provides a defend-in-place strategy, fire and building codes utilize a strategy of compartmentalization in order to reduce the likelihood of a fire spreading within a building. The concept is to divide the building into small "fire areas" or compartments in order to contain the fire within that given area, slowing its spread throughout the building, and allowing the occupants more time to escape. When the fire is contained within a smaller area, this also holds the fire in check until a responding fire department arrives at the building to commence extinguishing operations. Ultimately, compartmentalization contributes to life safety and property conservation by slowing a fire's spread and by containing the fire.

Ambulatory Surgery Center Safety Guidebook. https://doi.org/10.1016/B978-0-12-849889-7.00003-0

PURPOSE OF CONSTRUCTION FEATURES

In order to create these smaller fire areas or compartments, walls and other openings must be designed and constructed of materials that will prevent or retard the passage of heat, flame, and hot gasses. This is achieved by the codes requiring fire walls, fire barriers, fire doors, and other building materials that possess what codes refer to as "fire resistive ratings." Building materials and methods have achieved these ratings by the materials being tested by a recognized independent testing laboratory, which then assigns a fire resistive rating measured in minutes or hours, such as a 1-hour or 2-hour rating. The rating translates how long those materials and/or construction methods will withstand exposure to fire. For example, a 1-hour rating means that a wall constructed of those materials will remain intact after being directly exposed to heat and flames for 1 hour, at which point it will break down and the fire will spread through the materials. Fire resistive ratings are also assigned to construction components such as doors and windows that may be utilized in creating a compartment within a building.

Barriers Limit Smoke/Fire Spread

Because exposure to smoke is just as dangerous to a building's occupants as fire exposure, compartmentalization is also incorporated into design and construction to prohibit and slow down the travel of smoke throughout a building. In addition to fire-resistive walls, doors, and/or windows, codes may also require smoke barriers. Based on the square footage of a facility, it may need to be divided into sections called smoke compartments, in order to prevent the exposure of occupants to potentially dangerous smoke created by a fire in another part of the building.

Compartment Integrity

Compartmentalization is a key concept for ASC safety because it allows a "defend in place" strategy to be employed in order to protect the occupants. Employees of ASCs need to be aware of any required fire and/or smoke compartmentalization, because they will be responsible for ensuring compartments retain their integrity and will function as designed during a fire event.

TYPES OF FIRE AND SMOKE BARRIERS

Walls

Compartments are created by building a fire wall to subdivide a building. Fire walls will be assigned a fire resistive rating measured in minutes or hours.

The wall itself will extend continuously from the foundation or floor to the roof. It is important to understand that the visible interior ceiling is not usually the actual roof of the structure. Fire walls are constructed of a fire-resistive material, such as sheet rock, drywall, or concrete. All openings in a fire wall must be sealed in order to prevent the travel of smoke and fire beyond the fire wall, into any adjoining spaces. This can be achieved by using an approved (by fire and building codes) material to seal these openings that matches the fire-resistive time rating of the fire wall. Openings are typically sealed using fire caulk. Materials used must have the appropriate fire-resistive rating. In order to attain this rating, they will have an intumescent property, meaning that they will swell as a result of heat exposure. It is not uncommon for pipes and wires that pass through fire walls not to be sealed properly with a fire caulk material. This results in the fire wall not meeting its required fire resistance rating, causing it to be non-compliant with the code.

Most ASCs will have at least one fire wall. It is important that a facility's staff understand that maintaining the integrity of the fire wall is required by fire and life safety codes. Regular periodic inspections should be performed to inspect for any openings that may be present. This is especially important to do after any work has been performed above a dropped ceiling in a facility. A visual inspection should be carried out after work such as electrical, heating, or air conditioning is completed, or any time that new wiring or cables are installed for technology equipment. These projects often require breaching a fire wall to extend pipe, wires, or cables. When this occurs, the openings created need to be sealed with fire caulk around the newly installed objects that pass through the fire wall. This will create an airtight, and therefore fire-resistant, seal.

Doors
In order to maintain the integrity of the compartmentalization created by a fire wall, any openings in that wall, such as doors or windows, must possess a fire resistance rating matching that of the wall. This is also a code requirement. Doors required and installed in a fire wall are referred to as fire doors. The manufacturer of the fire door will obtain a rating for a given model of door that it sells, and the door and door frame will have a label affixed, noting that a third party testing laboratory attests that the door meets fire-resistive properties for a given period of time. Fire door ratings are expressed in minutes and hours. Fire-rated windows are similar. The glass and frame undergo testing and there will be a rating label (often etched or imprinted) on the glass surface, also expressed in minutes and hours.

ASC staff are responsible for the maintenance of fire doors and windows. It is important that the labels on these components are always left intact and readable so that a surveyor or inspector can read them to verify that the doors and windows are in fact tested and meet the required fire resistance ratings. Often, labels are removed or painted over, rendering them unreadable. Some fire doors will also have a rubber strip running along the edges to serve as a seal for smoke and hot gasses. This is part of the door that was in place when it was tested in order to receive the fire resistance rating, so this component must remain intact. Seals or gaskets on a fire door should never be removed, because this can change the fire resistance rating.

Fire doors can only perform their function of stopping the spread of fire when they are closed. This why fire doors are installed with automatic closing hardware. Therefore door stops should not be used to hold the door open. There is one exception, and that is if the door is equipped with an automatic hold-open device, consisting of magnetic fittings where one piece is affixed to the door and one piece mounted to the wall. The device is integrated with the building's fire alarm system, causing the door to release and close whenever the alarm is activated.

Roll-Down Automatic Fire Barriers

Occasionally a facility will be equipped with a fire door that rolls down either when exposed to the heat of a fire or when activated by the fire alarm system. This is called a roll-down fire door. Such doors are typically found in a window or pass-through area such as a reception desk. Sometimes they are used in place of a common hinged door, especially if there is a very wide doorway. Staff should be aware if the facility they work in is equipped with this variety of door, because it is critical that nothing is placed in the path of travel that would block the door as it is lowering, since this would render the door ineffective during a fire event.

Fire and Smoke Dampers

If a fire wall contains duct work that passes from one side of the wall to the other, the codes require the wall to be equipped with a device known as a fire or smoke damper. This is a device installed within the air ducts (heating and air conditioning) that is designed to close automatically upon detection of heat or smoke and therefore resist the passage of heat or smoke. Dampers thus maintain the integrity of the fire wall so that compartmentalization is ensured, protecting occupants from the passage of heat and smoke into other areas of a building. ASC staff should be aware if their facility contains dampers because the codes require inspection, testing, and maintenance of these devices.

INTERIOR FINISH BUILDING FEATURES

The codes also strive to limit the rate of fire spread in a building by regulating the type and quantity of materials used in the interior finish components. This is the exposed interior surfaces of a building that can either accelerate or prohibit the spread of fire and flame. One item within an ASC that are specifically regulated is curtains—specifically those used to ensure privacy between patient beds. Another interior finish item addressed by the codes applicable to an ASC is upholstered furniture. Both upholstered furniture and curtains must have labels in place that verify that they meet the flame-resistance characteristics prescribed by the codes.

OPERATIONAL REQUIREMENTS

The following building construction components are required to be inspected on a regular basis, and documentation of the inspections/maintenance needs to be retained on site and available for an inspector or surveyor to review:

- fire walls (recommended monthly);
- smoke barriers/walls on the perimeter of smoke compartments (recommended monthly);
- fire doors (recommended monthly);
- roll-down fire doors (annually); and
- fire/smoke dampers (annually).

CHAPTER 4

Means of Egress

IMPORTANCE OF EVACUATION

The earliest fire and life safety codes for buildings were based on the ability of a building's occupants to exit a building quickly and freely during an emergency. Generally the purpose of codes as they relate to exiting is to allow people to escape quickly enough to avoid a fire. Most fires involving great loss of life can be directly attributed to the compromise of a building's exits. Therefore, keeping the occupants safe from fire requires a quick, uninterrupted path to the outside of the building. In some larger health care environments, this may mean an area of safety such as an adjoining smoke compartment, as discussed in Chapter 3. The person exiting the building should not be required to utilize special tools or have existing knowledge of the building. The codes require the path to be clear, unobstructed, well-marked, and illuminated. ASCs must meet requirements for multiple parts or "components" of the egress or exit system. These components include the doors, corridors, signs, and lighting, all of which are critical to ensure that the occupant ultimately reaches the safety and sanctuary of the public way, which is the public sidewalk or street adjacent to the structure.

Components of the Exit Path

Probably the most obvious component of the exit path is the door. Doors must meet requirements for height, width, and the direction in which they swing. Minimum width requirements are addressed when a building is constructed to ensure the easy passage of occupants. Doors must also operate or swing in the appropriate direction, that is, towards the direction of egress, so that occupants exiting quickly do not have to pull it towards them, which might cause a delay. There are also requirements as to what type of hardware (mechanisms utilized to operate the door latch) can be installed on exit doors. Exit doors must be distinguishable from nearby construction materials so that they are easily recognized as doors. Reflective materials or finishes such as mirrors are not allowed. Additionally, doors cannot be covered by decorative items, draperies, and/or curtains.

Ambulatory Surgery Center Safety Guidebook. https://doi.org/10.1016/B978-0-12-849889-7.00004-2

Another main component of the exit system is a corridor. The codes define a corridor as an enclosed path of egress travel. Most people refer to this as a hallway. The purpose of these paths is to provide a clear, quick, and ideally protected route out of the building, and to connect exit doors which ultimately lead to a path out of the building. The codes define minimum widths of corridors and also set out requirements to eliminate potential obstructions within the corridors. Because some corridors are designed to protect exiting occupants from smoke and fire, they may need to be constructed of certain materials to limit the spread of smoke and fire. This type of corridor is referred to as "fire resistive" and the codes may require a certain level of fire resistance. Based on the degree of protection for occupants required, the codes will prescribe a level of protection defined as a "fire-resistance" rating. Once these corridors are designed and built into a structure, it is important that no breaches or openings are created in them, much like the fire walls and barriers described in Chapter 3.

The codes require exit doors to be marked and readily visible with signs that specifically say "EXIT." The codes define a specific size for the lettering: 6 in. tall with a stroke width minimum of ¾ in. Additionally, the codes require the paths of travel (exit corridors) and direction of the exit pathway to be indicated through the use of exit signs, especially when the path of travel towards an exit is not immediately obvious. The exit signs must include arrows indicating the correct direction of the exit path. Certain codes, which are applicable to ASCs, also require exit signs to be internally illuminated. The illumination of these signs must be continuous even during a power outage. The power to keep the sign illuminated may be supplied by batteries, built into the exit sign, or supplied from another type of external power supply such as a generator or battery backup UPS (uninterruptable power supply) system. ASCs must be equipped with exit signs that will remain illuminated for 90 min in case of loss of the facility's primary (normal) power supply.

The codes require all areas of an exit path to be illuminated whenever a building is occupied. Additionally, because there is potential for a power outage during a fire or other emergency, ASCs must provide emergency lighting that is powered by multiple sources. The secondary power that is supplied during an outage of the primary source may come from batteries or a facility's backup emergency generator or emergency battery system. The requirements for these systems are described in detail in Chapter 8. It is important to understand that emergency illumination is required throughout the interior exit pathway as well as on the building's exterior, so that the exit path is illuminated until the exiting occupant reaches the public way.

It is critical that all exit components are maintained in the state in which they were approved by the authority having jurisdiction. The exit path must be free from obstructions and must be instantly accessible in case of an emergency or fire whenever the building is occupied. This means that no equipment can be parked, placed, or stored within an exit corridor. It is also critical that no materials or supplies are stored with an exit corridor. Exit corridors must remain open and freely accessible at all times. It is important that staff are frequently reminded of these requirements and the necessity to maintain free and clear exit corridors and pathways.

OPERATIONAL REQUIREMENTS FOR EGRESS

The following components of the means of egress, or building exiting, are required to be inspected regularly, and documentation of the inspections/maintenance must be retained on site and available for an inspector or surveyor to review:

- exit signs in place and illuminated (weekly);
- exit pathways free of obstructions (weekly); and
- exit door hardware operates correctly (weekly).

CHAPTER 5

Portable Fire Extinguishers

WHAT PORTABLE FIRE EXTINGUISHERS DO AND WHY THEY ARE REQUIRED

The most common piece of fire protection equipment found in all types of buildings is the portable fire extinguisher. People are the most familiar with this equipment because fire and life safety codes require them to be in place in all types of buildings, and they are therefore seen every day. They are in the form of a red cylinder with a handle on top. Over the years, fire extinguishers have evolved in both the way they are used and in the types of extinguishing agent contained. An extinguishing agent is the product within the extinguisher that is applied to the fire.

The codes specify the location, type of extinguishing agent, maintenance, inspections, and training of staff relating to portable fire extinguishers within health care environments, and contain very specific requirements for an ASC. Once a building is designed, inspected, and approved for use as an ASC, the fire extinguishers must be inspected and maintained. Additionally, the facility must have documentation of staff being trained in the use of portable fire extinguishers.

The purpose of portable fire extinguishers is to provide a way for a building's occupants to suppress a fire in its initial stages, while it is still very small. This will contribute to the protection of occupants, especially when there are difficulties or delays associated with the evacuation of the occupants, such as may occur in a health care facility. Again, it is important to note that in order for portable fire extinguishers to be effective, personnel must be trained properly.

CODE REQUIREMENTS

All buildings are required by the codes to be equipped with portable fire extinguishers. The quick control of a small fire in order to protect

Ambulatory Surgery Center Safety Guidebook. https://doi.org/10.1016/B978-0-12-849889-7.00005-4

occupants who may have difficulties evacuating is especially applicable to health care facilities and ASCs.

Correct Type; Fire Rating Category

Because different materials have differing properties when they burn, such as wood compared to a flammable liquid, fire extinguishers must contain a relevant extinguishing agent to the type of fire that is most likely to occur where the extinguisher is located. Fire extinguishers are grouped or "rated" in four classifications to suppress four general categories of fires or fire hazards: Classes A, B, C, and D.

- Class A fire hazards generally include materials that are considered "ordinary combustibles" such as paper, wood, cloth, rubber, and most plastics.
- Class B fire hazards are materials are flammable and combustible liquids, and include alcohols, solvents, oil-based paints, and flammable gasses.
- Class C fire hazards include energized electrical equipment where it is critical that the extinguishing agent is nonconductive, so that that there is no danger of transferring an electrical current to the person operating the fire extinguisher.
- Class D fire hazards are fires involving flammable solids, the majority of which are combustible metals such as magnesium.

Most health care environments are classified as "light" or "ordinary" fire hazard areas. Therefore, the codes require a general-purpose fire extinguisher in these facilities. This type of fire extinguisher, which will suppress a combination of Class A, B, and C fires, or an "ABC," as they are referred to, is required in an ASC.

Mounted in Appropriate Location

In addition to the appropriate type of fire extinguisher in an ASC, the location in which extinguishers are mounted within the facility is essential to ensure adequate protection for the building and, ultimately, the occupants. In order that the extinguishers are readily accessible, the codes require a fire extinguisher to be mounted within 75 ft of any location with an ASC. The codes also include specific requirements as to how the extinguishers are to be mounted on the wall. Per the codes, fire extinguishers must be mounted so that the bottom is at least 4 in. above the floor and the top is no higher than 42 in. There are some exceptions that pertain to the size of the extinguisher, but these specifications also comply with accessibility requirements.

Accessible and Unobstructed

Portable fire extinguishers must also be mounted so that they are not hidden or obscured from view. If visual obstruction cannot be completely avoided, obvious signage needs to be provided to indicate the location. If extinguishers are located within a fire extinguisher cabinet, the cabinet must be unlocked or provided with a means of removal or access.

Training/Use

In addition to the availability of portable fire extinguishers, the codes require employees of health care facilities and ASCs to be familiar with their proper use. Facilities are required to maintain documentation of employee training, which commonly utilizes the PASS acronym to simplify the steps of employing a portable fire extinguisher to suppress a small fire:

- **P**ull the pin.
- **A**im the extinguisher at the fire.
- **S**queeze the handle to release the suppressing agent.
- **S**weep the spray pattern back and forth at the base of the fire.

Training in the use of fire extinguishers can be obtained from several sources. One of these is local fire departments. Many fire departments' Fire Marshals' Offices or Fire Prevention Bureaus offer classes, often at no cost or for a nominal fee. Another resource is vendors that provide fire extinguishers and maintenance of fire extinguishers. Many will include training sessions as part of their initial supply and/or maintenance contract. On completing the training of staff members, documentation must be retained to record the training.

OPERATIONAL REQUIREMENTS FOR FIRE EXTINGUISHERS

Because they need to be readily available and absolutely must work properly when needed, the codes require regular maintenance and inspections of fire extinguishers. Some of these tasks can be performed and satisfy the codes by the facility staff, such as 30-day checks to ensure that a fire extinguisher is in its proper location, checking that the gauge indicates that the extinguisher is full, and simply checking that the fire extinguisher is accessible and unobstructed. However, the code also requires that an annual inspection and maintenance are performed by a qualified technician, who is trained and certified in fire extinguisher maintenance. Records must be kept of all inspections and maintenance performed, whether these are completed by facility staff or by an external service provider.

OPERATIONAL REQUIREMENTS FOR EGRESS

The following inspections and maintenance are required for portable fire extinguishers, and documentation of the inspections/maintenance must be retained on site and available for an inspector or surveyor to review:

- visual inspection of proper placement and operational readiness (weekly); and
- third party annual inspection and maintenance.

Fire Alarm Systems

PURPOSE

One of the main objectives of codes is to ensure that a building's occupants are able to evacuate safely. The purpose of a fire alarm system is to initiate the evacuation process by providing notification that a fire has been identified, as early as possible. The alarm system, when activated, provides both audible and visual signals to inform the occupants that evacuation is necessary. Additionally, the fire alarm panel will also provide notification, via phone lines, to an outside monitoring agency, who will notify the appropriate fire response agency of a potential fire or emergency situation, who will then respond. The fire and life safety codes require fire alarm systems in certain buildings and/or parts of buildings based on the activities of the occupants who inhabit or occupy that building. Since health care facilities, and ASCs in particular, are occupied by people who may have a delayed and/or inhibited ability to evacuate, the codes require these facilities to be equipped with a fire alarm system.

TYPES OF FIRE ALARM SYSTEMS

Fire alarm systems are divided into two basic categories, according to how they are initiated or activated: automatic or manual. An ASC is required to be equipped with one or the other, depending on the facility's size and level of anesthesia administered. When there is greater risk to occupants and patients, there is a greater likelihood that an automatic system will be required.

Automatic

Automatic fire alarm systems utilize an automatic means to initiate the alarm components of the system, just as the name implies. The activation of this type of alarm systems requires no human action. This is achieved by utilizing devices such as smoke detectors, heat detectors, flame detectors, and fire sprinkler water flow switches. The activation of one of these

Ambulatory Surgery Center Safety Guidebook. https://doi.org/10.1016/B978-0-12-849889-7.00006-6

devices will initiate the audible and visual warning devices, as well as the alarm panel sending signals to the monitoring agency in order to initiate a fire department response.

Manual

Manual fire alarm systems require action or input by the occupants of the building in order to activate the alarm system, including the visual and audible warning devices. These systems are simpler than automatic systems and are comprised of manual pull stations, the fire alarm panel, and the audible and visual warning devices. The activation of one of the manual pull stations will initiate the warning devices as well as sending a signal to the monitoring station that the system is in alarm mode, requiring notification of the fire response agency.

COMPONENTS

Fire and life safety codes require the components of fire alarm systems installed in ASCs to meet certain requirements. One requirement is that they are approved for use in the system by meeting the testing requirements of a recognized major testing or listing agency. Another requirement is that the various components are compatible to be used together. There are specific code requirements that address how and where specific components are installed.

Fire Alarm Panel

The alarm panel is the "brains" of a fire alarm system. It provides information on the status of the system and provides power to the various components. When an input or initiating device, such as a pull station, smoke detector, or fire sprinkler water flow switch, is activated, the panel will switch on the audible and visual warning devices. Additionally, the panel will send signals, via phone or Internet, to the alarm monitoring agency. Fire alarms panels are often referred to as "Fire Alarm Control Panels" (FACP).

Manual Pull Stations

Manual pull stations provide the occupants of a building with a readily identifiable device to activate the fire alarm system in a building. The device is required to be red in color and to indicate with contrasting white lettering that it is a fire alarm activation device. Codes specify where manual pull stations must be located. Typically they are in exit paths near each required exit. Manual pull stations are activated by an individual pulling down on a designated lever or handle. Activation of a manual pull station will initiate the fire alarm warning devices so that a building's occupants know there is a fire or other emergency requiring evacuation of the building. ASCs are required to have manual pull stations installed at each required and designated exit.

Smoke and Heat Detectors

A fire alarm system that is automatic and requires no human activation for the system to go into alarm mode and initiate the evacuation signals must be activated by some type of detector. Detectors sense either smoke or heat. Heat detectors are designed to send an alarm signal to the fire alarm panel if the ambient temperature reaches a predetermined level, usually 135–165°F. Smoke detectors, as the name implies, are designed to identify a fire while in its early flame or smoldering stages. When the device's internal equipment senses smoke, the device sends an alarm signal to the fire alarm panel.

Horn/Strobe Devices

After alarm signals are sent to the fire alarm panel, the panel must notify the building's occupants that an emergency is underway. This is achieved by the activation of the alarm system's notification devices. Fire and life safety codes require notification to be both audible and visual. To attract the attention of a building's occupants, audible alarms must be distinctive and recognizable. Because some occupants may be hearing impaired, the code also requires a visual signal to be produced by a strobe light. It is important that facilities contain the appropriate number of audible and visual devices so that occupants at any location within the facility will see and hear a fire alarm when activated. ASC staff should keep this in mind and ensure that fire alarm signaling devices are not obscured by anything, such as privacy curtains or storage.

Monitoring of Alarms Outside an ASC

Fire alarm systems are required by fire and life safety codes to be monitored off site. The code has specific requirements that monitoring agencies or facilities must meet to be considered "approved." Monitoring facilities/agencies must be operated and maintained by personnel whose primary business is to furnish, maintain, record, and supervise a signaling system. Monitoring of the system means that a person watches what the system status is 24 h a day, 365 days a year. If the system goes into alarm, the appropriate fire response agencies will be contacted and they will respond to the facility, with the assumption that a fire is occurring. ASC facility staff should always have the monitoring agency's phone number and their facility's account number readily available in case they need to contact the agency for problems or service needs, and especially so they can inform the monitoring agency whenever the ASC conducts a fire drill.

An ASC can use the fire alarm system to simulate an actual fire alarm, to include actually activating the alarm by using one of the alarms manual pull station. This will cause the notification devices to go into alarm mode, giving the staff an understanding of what the alarms look and sound like so that they know what to expect during an actual fire alarm situation. The monitoring agency will need to be contacted prior to the fire drill and given instructions that the ASC will be conducting a drill. The monitoring agency will then ensure that the fire alarm system is in "test mode" and no emergency responders will be dispatched.

Fire alarm panels are designed to provide constant information on the status of the fire alarm system. The most obvious condition is when the system is in alarm. The alarm panel will also provide information if something is not working properly in the system. There are two conditions, other than alarm, with which the ASC staff should be familiar: "supervisory" and "trouble." Both conditions will cause the fire alarm panel to emit an audible beeping sound. Additionally, the monitoring agency will call the contact person listed on the account and advise them of the alarm sensing a potential problem requiring attention if that person is off site and not hearing the panel's audible indicator. These conditions are both intended to let the building's occupants know that the fire alarm requires the attention of a qualified technician to service the system and correct whatever problem the panel is sensing outside of normal parameters.

Code Requirements for ASCs

The fire and life safety codes require all ASCs to be equipped with a fire alarm system with manual pull stations and audible and visual warning devices. Some facilities, depending on size and other factors, may be required to have additional components such as smoke and/or heat detectors.

The codes require regular inspection, testing, and maintenance of the system. It is important that the staff ensure that these tasks are completed at the required frequency and that documentation is maintained. The fire alarm panel must always be visible and accessible, and not obstructed by storage or decorations.

If the panel is not readily visible, the door leading to its location must be marked with "Fire Alarm Control Panel" or "FACP," so that emergency responders can locate it quickly to obtain information about what is occurring at the facility.

All ASCs must be equipped with manual pull stations. The fire and life safety codes contain very specific requirements on the locations of these. In an ASC, a manual pull station must be installed within 5 ft of each designated exit from the facility. The device must be installed on the interior side of the exit door, so that an exiting occupant can activate the device prior to going through the exit door.

Manual pull stations must be visible and readily accessible at all times. It is important that no objects or decorations are placed where they could inhibit or delay access to a manual pull station.

As described earlier, manual pull stations can be utilized for drills; in fact the fire and life safety codes require them to be used for at least one fire drill per year. However, prior to the use of the stations, the monitoring agency must be contacted so that when they are activated, emergency responders are not sent to the facility.

OPERATIONAL REQUIREMENTS FOR FIRE ALARM SYSTEMS

The following inspections and maintenance are required for fire alarm systems. Documentation of the inspections/maintenance must be retained on site and available for an inspector or surveyor to review:

- control equipment/control panel (weekly);
- batteries in control panel (monthly);
- water flow actuator devices/flow switches, battery charge, control equipment, supervisory signal devices, signal transmission (quarterly, professionally recommended); and
- overall system inspection, testing, maintenance (annually).

CHAPTER 7

Fire Sprinklers

PURPOSE

Built-in fire protection systems provide property and life safety protection that is considered "active" by fire and life safety codes. Active fire protection systems are present in a building but are only initiated, or become "active," when there is an actual fire. Fire sprinklers are a type of "active" fire protection system required by fire and life safety codes. Most ambulatory surgery centers will be protected by fire sprinkler systems. However, some exceptions in the codes allow certain smaller facilities to exist without fire sprinklers.

Fire sprinklers are an integrated system of underground and overhead piping and a connected water supply. The system is activated by heat from a fire, at which time water will be discharged over the fire from the piping system; the water flows from a nozzle, referred to as the fire sprinkler head. The system is connected to a reliable water supply to ensure that there is always water ready to extinguish a fire. The piping overhead in the building is a network of specially sized and hydraulically designed piping. Valves are located at the point in which the supplying water pipe enters the building, called a fire sprinkler riser, so that the system can be controlled and turned on or off. There is also a device located at the fire sprinkler riser; this senses when water is flowing and activates the fire alarm system.

It is important to remember that fire sprinklers only discharge water when exposed to heat produced by a fire. The sprinkler heads are designed to activate at a certain predetermined temperature through the melting of a metal device called a fusible link or a liquid-filled glass bulb breaking. When this occurs, the failure of these objects creates a void and allows the pressure of the water in the pipes to push open a plug, so that water flows out of the system. Only those fire sprinkler heads exposed to the heat of a fire will activate.

Ambulatory Surgery Center Safety Guidebook. https://doi.org/10.1016/B978-0-12-849889-7.00007-8

COMMON MISCONCEPTIONS/MYTHS

A common misconception is that all fire sprinklers in a building will begin emitting water when there is a fire, smoke, and/or when the fire alarm is activated. This is not true, and is a myth that has been perpetuated by TV and movies. Only fire sprinkler heads that are exposed to heat will actually discharge water. Fire sprinklers further away from a fire will only discharge water if they are exposed to enough heat to reach their predetermined discharge temperature. Interestingly, statistics from actual fire incidents show that typically, only one or two fire sprinkler heads will extinguish or hold a fire in check, preventing it from spreading, until the responding fire department arrives.

CODE REQUIREMENTS

When a fire sprinkler system is designed and installed, the codes require all rooms and spaces to be fully protected by fire sprinkler heads, by ensuring that the water spray will cover all the floor area in a room and ultimately the entire building. A fire igniting in any location will thus be extinguished. The codes also specify how close any object may be placed to a fire sprinkler head without that object altering or obstructing the sprinkler head's spray pattern and affecting its fire extinguishing performance.

The occupants of ASCs need to be aware that the codes do not allow objects of any kind to be stored within 18 in. of a fire sprinkler head. This is usually not a problem in areas other than storage rooms. A common method of reminding facility staff to keep storage 18 in. below fire sprinklers at ceiling level is to mark the wall with a line (either painted or utilizing colored masking tape) 18 in. below the ceiling/fire sprinkler level, around the perimeter of the room. Occupants then have a constant visual reminder of the storage height limitation within that room.

OVERVIEW OF ADMINISTRATIVE REQUIREMENTS

Fire and life safety codes state that owners and occupants of ASCs are legally responsible for maintaining fire sprinkler systems and keeping the systems in good operating condition. Specifically, systems must be inspected, tested, and maintained according to NFPA 25, which defines the requirements. The code requires many of these tasks to be performed by a qualified technician on an annual basis. The vast majority of ASCs will not have a qualified technician on their staff, so it is recommended that the facility contracts with a third party fire sprinkler company. The annual inspection and test includes an examination of components such as sprinkler heads, valves, pipes, and pressure gauges.

It is critical that facilities maintain all written documentation relating to inspection, testing, and maintenance from the time the system was installed (when the ASC was being constructed). During CMS (Centers for Medicaid and Medicare Services) surveys, the surveyor will ask to review the inspection testing and maintenance records of the facility's fire sprinkler system.

OPERATIONAL REQUIREMENTS FOR FIRE SPRINKLERS

The following inspections and maintenance are required for fire sprinkler systems, and documentation of the inspections/maintenance must be retained on site and available for an inspector or surveyor to review:

- system gauges, control valves, fire pump casing and pressure relief valves, and backflow prevention reduced pressure valves and detectors (weekly);
- wet and dry gauges, control valve locks, control valve tamper switches, exterior alarm devices, exterior dry pipe valves, and exterior quick-opening devices (monthly);
- alarm devices, hydraulic name plate, sprinkler system valves, and fire department connection valves (quarterly); and
- overall system inspection, testing, and maintenance (annually, third party qualified technician required).

CHAPTER 8

Emergency Power, Supplies, and Electrical Systems

PURPOSE

Fire and life safety codes contain many requirements for electrical components in all types of buildings. Because of the critical life support role of electrical systems in health care facilities, these facilities must comply with even more rigorous requirements. There is an expectation that the systems will be safe and reliable, and during a power failure, there will be a quick, almost unperceivable, restoration of power. There are myriad codes to ensure highly dependable and safe electrical service in ambulatory surgical centers (ASCs). The codes define what type of equipment is required and how it is installed, and contain specific requirements applicable to the components' inspection, testing, and maintenance. Most of the code requirements are based on the fact that ASCs contain sophisticated medical equipment that requires a reliable source of electrical power to perform and will continue during an electrical utility source failure.

Code References

Because of the importance of a reliable, constant delivery of power to equipment in an ASC, the fire and life safety codes require ambulatory surgery centers to be equipped with an emergency power supply. NFPA 99, which defers to NFPA 110 and NFPA 111, specifically addresses and defines the type of system that a facility must have in place. The codes consider what level of anesthesia will be administered within a facility to determine what level of sophistication an emergency power supply will need to possess. The more intense the level of sedation, the more sophisticated the level of emergency power system required. The codes require a less sophisticated, or Type III, emergency power system for all levels of sedation other than general anesthesia. When general anesthesia is administered, the codes require a Type I emergency power system. Type I systems are more complex and have a number of redundancies built into them to protect a patient who is being treated with general anesthesia, is unconscious, and is reliant on medical equipment to sustain life.

Ambulatory Surgery Center Safety Guidebook. https://doi.org/10.1016/B978-0-12-849889-7.00008-X

In addition to a facility being equipped with the proper type of emergency power system, the codes contain specific requirements for the inspection, testing, and maintenance of emergency power systems. All of these requirements should ensure that these systems will perform as intended during a loss or interruption of the normal utility power service.

TYPES OF SYSTEMS

There are two types of power sources for emergency power systems installed in ASCs that will provide power in the facilities. The most common is a generator. The other is a battery-based or "stored" power system called an uninterruptable power supply (UPS). Both types of power source are integrated into the facility's electrical system so that when a power loss is detected, they will automatically initiate the delivery of power to the essential portions of the electrical system requiring power for patient and other occupant safety.

Generators

Generators are the most common source of power for emergency power systems in ASCs. They are powered by either diesel fuel or natural gas. The size of the generators will vary based on the demand of the facility, and will be selected by the architects and engineers when the facility is built. Like any piece of equipment, generators require regular inspections, testing, and maintenance; these tasks are prescribed and defined in the codes for an ASC.

Battery/Uninterruptable Power Supply

Another source of power for emergency power systems in ASCs is a stored power system, which in an emergency provides power derived from a system of batteries. One advantage of this system is that the switch from a normal power supply to power from the stored system is nearly immediate and almost imperceptible. As with a generator, stored power systems must be inspected, tested, and maintained as outlined in the fire and life safety codes.

SPECIFIC CODE REQUIREMENTS

In order for an ASC to comply with current codes, the facility must meet a number of requirements for the installation, inspecting, testing, and maintenance of electrical components and systems.

System Components

The installation of generators is one area that is reviewed by compliance surveyors when conducting an ASC site visit. Many of the emergency power and electrical system components will have been installed when a facility was built; however, the surveyor will confirm if the required components are in place and meet code requirements.

The emergency power system is required to have a power unit component; this will be either a generator or a stored power system of batteries. The generator must be mounted either within a secure enclosure or in a cabinet. The enclosure needs to contain a battery-powered light, so that a service person would have light to service the generator, in the case of a mechanical failure. Generators are also required to be equipped with an emergency shutdown switch, so that the generator may be immediately shut down if required in an emergency. All generators must also be equipped with a separate device called a transfer switch; codes require this to be readily accessible. A transfer switch is a device that senses a loss of power, and provides a signal to the generator to start supplying power to the facility. Another code requirement for generators is a device called a remote annunciator, which is a status panel that provides information about the generator to the occupants of the ASC. This must be installed inside the facility in a location that is observable by the staff of the ASC, so that they are aware of information pertinent to the generator and are aware of its status and readiness.

Additional Requirements for General Anesthesia Locations

Facilities that administer general anesthesia must also meet more extensive requirements in the areas of emergency power and electrical systems. Most of the requirements will have been addressed when a facility is constructed, but some aspects will need to be maintained by the facility's occupants. One of these requirements is that within operating rooms, some type of battery-powered lighting is installed that is constantly charging under normal power conditions. This is a virtually redundant system that ensures lighting, which will be supplied immediately, during a transfer of power from the normal source to the generator. A generator requires a few seconds for the engine to start, and thus to begin supplying power. The battery-powered light will cover that potential gap in service. These lights can be either surface-mounted or integrated into the light fixtures, and must be part of the inspection, testing, and maintenance program.

Medical Facility Power Outlets/Breakers

Because continuous and uninterrupted electrical power for medical equipment is such a critical component of patient care and life support, the fire and life safety codes contain specific requirements describing the number of electrical outlets installed in a facility, the labeling of circuit breakers and electrical panels, as well as the inspection, testing, and maintenance of these pieces of equipment.

Electrical outlets in medical facilities must meet certain requirements for durability to ensure that there is enough tension to hold a plug in place so that it cannot be accidentally knocked out or fall out of the receptacle, potentially rendering a critical piece of medical equipment out of service. The codes require an ASC to conduct annual tension testing of outlets and electrical breakers to ensure reliability for medical service.

OPERATIONAL REQUIREMENTS FOR EMERGENCY POWER SYSTEMS

The following inspections and maintenance are required for emergency power systems, and documentation of the inspections/maintenance must be retained on site and available for an inspector or surveyor to review:

- On a weekly basis (at intervals of not more than 7 days), the generator set will be inspected for compliance with NFPA 110 requirements, including:
  - storage batteries used in connection with essential electrical system generator sets are maintained in full compliance with the manufacturer's specifications; and
  - all accessory components are maintained in full compliance with the manufacturer's specifications.
- Generator sets are tested under load conditions 12 times a year (at intervals of 20–40 days).
- Transfer switch(es) are electrically operated from the standard position to the alternate position, and then returned to the standard position.

The following inspections and maintenance are required for battery-powered emergency power systems:

- Battery electrolyte levels are checked, terminals and intercell connectors are cleaned and regreased (if necessary), and cell tops are cleaned. Individual cell voltages are checked. The specific gravity of pilot cells is checked and recorded, where applicable. The condition of the plates and sediment in free-electrolyte lead-acid batteries with transparent containers is noted (monthly).

- Equipment is exercised under a connected load for a minimum of 5 min (quarterly).
- The battery system is tested at full load for the full EES duration (annually, by a qualified technician).

CHAPTER *9*

Medical Gases

RISKS AND HAZARDS OF MEDICAL GASSES

Ambulatory surgical centers utilize a variety of pressurized medical gasses, ranging from those contained within small portable to cylinders to larger cylinders stored within designated rooms connected to pre-piped systems that deliver the gas to various rooms with the facility. Compressed and pressurized gasses and associated equipment can present fire and other life safety hazards to occupants of ASCs; therefore the fire and life safety codes contain requirements for the presence and use of gasses within these facilities.

One medical gas that is a fire safety concern is oxygen. As a single element, oxygen is not flammable; however, it is a catalyst to combustion and greatly accelerates the combustion process, being an oxidizer. Additionally, some types of medical gasses are considered "inert," meaning that they are not flammable in themselves, nor are they a catalyst to combustion. However, they are a life safety hazard because if there were an accidental release or leak that would accumulate in a small space, the ambient atmospheric oxygen in the room could be displaced and create a suffocation hazard for people within that space. The codes address these risks to protect both the patients and staff in a facility.

TYPES OF SYSTEMS

Based on the procedures performed within an ASC, some facilities will contain oxygen only in small portable cylinders. These cylinders may be used only occasionally or even may only be in place in case of an emergency situation where a patient may require oxygen. Facilities that perform more complex procedures and/or surgeries may contain larger medical systems and distribution systems to include medical gas storage rooms, distribution systems, and monitoring equipment. The more complex the system, the more code requirements a facility must adhere to in order to ensure compliance.

Ambulatory Surgery Center Safety Guidebook. https://doi.org/10.1016/B978-0-12-849889-7.00009-1

CODE REQUIREMENTS

The fire and life safety code requirements for medical gasses in an ASC are found in NFPA 99, "Standard for Heath Care Facilities." Because medical gasses are such a specialized hazard, the code specifically addresses the concerns and hazards for a health care facility. All piped medical gas systems must be inspected and tested by a technician who is qualified to inspect and certify piped medical gas systems. It is important that ASCs maintain all written records of these inspections and testing, as well as any maintenance that is performed on the system and its components.

Portable Cylinders

Because of the hazards and risks associated with medical gasses that are compressed and stored within portable cylinders, facilities must ensure they adhere to the following requirements for safety. All portable cylinders must always be secured to prevent them from tipping over or being damaged. This can be achieved by always ensuring that portable cylinders in use are within either a rack or a stand, with the proper regulator securely in place. This is because the cylinders contain medical gas that is being stored under high pressure, sometimes as high as 3000 lb per square inch. If a cylinder were to be accidentally knocked over and damaged, an accidental release at this high pressure could cause the cylinder to be propelled at a high speed, creating an unintentional missile that could violently damage anything it its path. Another requirement is that cylinders are not placed so that they obstruct a corridor or passage width.

Fire and life safety codes also limit the amount of portable cylinders in use within an ASC space. Codes allow up to 300 cubic feet of gas within an ASC smoke compartment. If an ASC is not divided into smoke compartments, 300 cubic feet may be the maximum for the entire facility. As a point of reference, small "E" size cylinders contain approximately 23 cubic feet of oxygen. A good rule of thumb is to keep the total number of small cylinders outside a designated medical gas storage room to 12 maximum. This is why commercially purchased storage racks that hold "E" cylinders will only accommodate 12 cylinders.

Piped Systems

Piped medical gas systems are built-in systems designed to store and distribute medical gas to patients while preventing any contaminants from entering the supply and also preventing any type of leak leading to a safety concern.

Piped medical gas systems contain four major components: source of the gas; distribution mechanism; outlets; and alarm and monitors.

Piped medical gas systems must be monitored through a panel that will provide information on the status of the gas levels in the system and if a supply is getting low. The panel will provide an audible alarm if the system's status falls outside of normal parameters. The codes require the facility staff to conduct regular checks of the monitoring panels to ensure these are operating correctly.

Because it is important for the staff to have the ability to shut off the medical gas supply quickly to outlets in the system, in case of an emergency such as a leak or a fire, piped medical gas systems must be equipped with emergency shut-off valves throughout the system. The codes require that these emergency valves are readily accessible at all times and that they are not obstructed by equipment or storage of materials. The valves must also be labeled to indicate what portion of the system they serve.

ASCs that are equipped with piped medical gas systems will have a designated room or closet that contains the medical gas supply cylinders, valves, and manifolds that are the entry point for the medical gas into the facility's piping system. The medical gas room is required to have signage on the entry door to identify it as the medical gas storage room, because of the potential hazards associated with storing medical gas cylinders. The codes require the cylinders to be individually secured to prevent them being tipped over. This is typically achieved by wrapping a chain around each cylinder and securing the chain to the wall. Because the large cylinders utilized in piped medical gas systems are steel and heavy, and can therefore potentially damage light switches and wall outlets, the codes require these electrical components to be mounted at least 60 in. above the floor in medical gas storage rooms. This prevents the light switches and outlets from being damaged by the cylinders when they are being exchanged. A damaged light switch or outlet that is energized has the potential to cause an electrical spark and subsequent fire. Larger medical gas storage rooms must be equipped with ventilation openings and possibly fans. The quantity of gas stored within will dictate whether or not a room has to be equipped with these features, and they will be installed at the time of construction of the facility. If these features are installed, it is important that they are maintained so that they continue to operate as intended. The fire and life safety codes also require the valves and manifolds of medical gas piping to be labeled to indicate the type of gas they will control.

OPERATIONAL REQUIREMENTS FOR MEDICAL GAS SYSTEMS

The fire and life safety codes require piped medical gas systems to be "certified" through inspection and testing by a qualified technician when the system is installed in a facility. Additionally, any time that maintenance is performed on the system or its components, the system must be inspected and tested. Medical gas system alarm panels are also required to be tested and the testing documented at regular intervals.

Inspection, testing and maintenance by a qualified technician (annually) is required for medical gas systems, and documentation of the inspections/maintenance must be retained on site and available for an inspector or surveyor to review.

CHAPTER *10*

Medical Equipment

RISKS AND HAZARDS

Because most medical equipment used for patient diagnosis and treatment is powered by electricity, there is always a concern for safety due to the risk of electrical shock. The fire and life safety codes contain requirements pertaining to the selection and use of medical equipment utilized in an ambulatory surgery center. One of these requirements is that all cord-connected electrically powered medical equipment that will be used in the vicinity of patients that are not double-insulated, that is, designed so that the casing will not transfer an electrical shock to the user should an internal wire become disconnected or loose. Additionally, the equipment is equipped with a three-wire power cord that includes a three-pin grounding type plug. The codes also require all medical equipment to undergo inspection and testing by a qualified technician. This is often referred to as "bio medical" testing.

CODE REQUIREMENTS

Generally, the codes require equipment to be tested at least once a year. However, equipment that is used in "wet" locations and critical care areas should be tested at least every 6 months. The testing of the equipment takes two forms, including:

- general electrical safety testing that takes place on all equipment
- specific testing for each piece or type of equipment, which may include functional testing and calibration.

If a facility has an operating room equipped with line isolation monitoring equipment, the fire and life safety codes require this equipment to undergo annual testing and calibration as well, even though it is installed and may be considered part of a building's overall electrical system.

Ambulatory Surgery Center Safety Guidebook. https://doi.org/10.1016/B978-0-12-849889-7.00010-8

OVERVIEW OF ADMINISTRATIVE REQUIREMENTS

As with all tests, it is important that a facility maintain proper records and documentation. When a qualified technician completes inspection and testing, they will typically tag each piece of equipment with a small sticker indicating who completed the test and the date it was performed. Additionally, they will provide a completed inventory documenting the inspection and testing of the equipment. Inspectors and surveyors will look at equipment and request inspection and testing records for their review, to ensure code compliance.

OPERATIONAL REQUIREMENTS FOR MEDICAL EQUIPMENT

Medical equipment is required to be tested by qualified technician prior to being placed into service, following any repair or modification, and then at the following intervals. Documentation of the inspections/maintenance must be retained on site and available for an inspector or surveyor to review, relating to:

- general care areas (12 months);
- critical care areas (6 months); and
- wet locations (6 months).

CHAPTER 11

Alcohol-Based Hand Rub Dispensers

RISKS AND HAZARDS

Alcohol-based hand rubs are essential for controlling the spread of infectious organisms in ambulatory surgery centers. Studies on their use have found that they are vastly more effective against numerous pathogens and are more effective for a longer period of time compared to hand washing with soap and water.

Alcohol-based hand rub is typically packaged in pump bottles or soft bladders for insertion into pump dispensers. The alcohol content may be up to 95%, and alcohol is classified as a Class I flammable liquid, which means that it is highly flammable. Soon after the benefits of alcohol-based hand rub were recognized and their subsequent popularity in health care facilities, fire safety officials became concerned with the related fire safety risks. The American Society of Healthcare Engineers commissioned a fire modeling analysis to study the overall effects of placing dispensers in corridors and suites of rooms. The results showed that with adequate spacing and other installation criteria, alcohol-based hand rubs could be utilized in health care environments with a reasonable degree of safety.

The results of these fire modeling scenarios and those conducted by other engineering groups are the basis for the requirements found in the fire and life safety codes that apply to ASCs. These include: the minimum distance that hand rub dispensers are required to be from each other; not mounting a dispenser over a potential ignition source; and the maximum quantity of hand rub dispensers that may be mounted within a given area within a health care facility.

CODE REQUIREMENTS

Dispenser Mounting

In order for a facility to comply with fire and life safety codes, it must adhere to the following requirements when mounting alcohol-based hand rub dispensers on the walls of an ASC.

Ambulatory Surgery Center Safety Guidebook. https://doi.org/10.1016/B978-0-12-849889-7.00011-X

- Any corridor a dispenser is installed within must be at least 6 ft wide.
- Multiple dispensers must be installed at least 4 ft apart.
- Alcohol-based hand rub dispensers cannot be installed directly over or directly adjacent to an ignition source such as a light switch or electrical outlet.
- Dispensers can only be installed in carpeted facilities if a fire sprinkler system is also in place within the building.
- Staff members must ensure that dispensers are maintained in accordance with the manufacturer's specifications.

Quantity Limitations

There are also restrictions regarding the quantity of alcohol-based hand rub that can be used within a facility. This is because it is classified as a Class I flammable liquid by the codes; therefore the greater the quantity, the higher the risk to the facility and its occupants. The maximum dispenser fluid capacity must not exceed 0.3 gal (1.2 L) if dispensers are installed in rooms and areas open to corridors. If they are installed within suites of rooms, the maximum fluid capacity must not exceed 0.5 gal (2.0 L). When multiple dispensers are installed within a single smoke compartment, the total combined quantity must not exceed 10 gal.

OPERATIONAL REQUIREMENTS FOR ALCOHOL-BASED HAND RUB DISPENSERS

Monthly visual checks for proper installation are required for alcohol-based hand rub dispensers, and documentation of the inspections/maintenance must be retained on site and available for an inspector or surveyor to review.

CHAPTER *12*

Facility Policy Requirements

PURPOSE AND TOTAL CONCEPT

Because the occupants of ambulatory surgery centers (ASCs) have varying degrees of physical ability to evacuate during a fire or other emergency, the codes contain requirements for facilities to have policies and procedures in place to increase the occupants' chances of survival. These requirements outline exactly what policies and procedures must address and include requirements for facility and equipment maintenance and also for policies that ensure the staff will know what to do in an emergency and therefore react properly. The intent is to have a complete system in place, or a "total concept," providing the safest environment possible for the patients and staff of an ASC.

CODE REQUIREMENTS

Facility Maintenance

In order to guarantee that building fire safety features and other equipment will function properly during an emergency event, fire and life safety codes contain requirements for regular inspection, testing, and maintenance. Each system or feature has unique requirements specific to that particular system, feature, or piece of equipment.

Fire Alarm System

In an ASC, the purpose of the fire alarm system is to identify a fire in its initial development stages and to provide an early notification to the occupants of the facility. The codes require facilities to maintain documentation of regular inspection and testing, as well as documentation of the design of the system, the manufacturer's operating instructions, and a document completed by the installing contractor referred to as the "Certificate of Completion." A facility's policies and procedures should note that the facility will meet all code requirements for inspection, testing, and maintenance, specifically an annual inspection and test to be performed by a "qualified technician" as

Ambulatory Surgery Center Safety Guidebook. https://doi.org/10.1016/B978-0-12-849889-7.00012-1

set out in the codes. In order to demonstrate code compliance to inspectors or surveyors, it is important that facilities maintain records of all fire alarm system inspection, testing, and maintenance.

Fire Sprinkler System

Fire sprinkler systems are one of the most important fire protection features incorporated into the total concept of protecting occupants of an ASC from fire. Statistically, the only times that fire sprinkler systems have not controlled a fire in a building are when they have not been maintained properly. If an ASC is equipped with fire sprinklers, the facility is required to maintain documentation of the installation, inspection, testing, and maintenance of the system. This includes any documents relating to when the system was installed—typically at the time the building in which the facility is located was built. Fire and life safety codes also require the facility to incorporate policies and procedures relative to the ongoing inspection, testing, and maintenance of the system to include an annual inspection and test by a qualified technician. These documents and records must be on site during an inspection or survey of the facility.

Laboratory Safety

Although the majority of ASCs do not contain a laboratory as defined by the fire and life safety codes, if a facility does contain a laboratory, it must comply with all code requirements, including policies and procedures for safety. These requirements are found in NFPA 45 "Standard for Fire Protection For Laboratories Using Chemicals," which includes requirements for electrical systems, emergency showers, and storage and use of flammable and combustible liquids. An ASC's policies and procedures must reflect the requirements in NFPA 45; in particular, the facility's fire and emergency plan should include a portion that is specific to the laboratory.

Emergency Power System

Because the survival of many patients being treated in ambulatory surgical centers depends on medical electrical equipment operating uninterrupted, fire and life safety codes require regular inspection, testing, and maintenance of the emergency power system. Facilities must retain all testing documentation performed on emergency power systems when they are installed. Additionally, the codes require periodic inspection and testing at monthly and annual intervals. Whenever maintenance is performed on these systems, the documentation must be retained. Review of all inspection, testing, and maintenance records is part of the inspection and survey process to determine code compliance. The policies and procedures of a facility need to include inspection, testing, and maintenance procedures that reflect code requirements to demonstrate facility compliance. Facilities equipped with emergency power systems that are supplied by a battery-type system must comply with NFPA 111 inspection, testing, and maintenance requirements. Facilities utilizing a generator to supply emergency power must comply with NFPA 110's requirements for inspection, testing, and maintenance.

Portable Fire Extinguishers
Portable fire extinguishers have proven to have been an effective means to control fires for decades. Because of this effectiveness, the codes have very specific requirements for inspection, testing, and maintenance, and ASC facilities must include these requirements in the facility's policies and procedures. The fire and life safety codes require that portable fire extinguishers are checked monthly, which can be done by facility staff, and annual inspection, testing, and maintenance tasks must be performed by a qualified technician. All requirements for portable fire extinguishers in ASCs can be found in NFPA 10, and this document should be referenced when developing facility policies and procedures pertaining to inspection, testing, and maintenance of portable fire extinguishers.

Building Heating, Ventilation, and Air Conditioning Systems
In order to ensure occupants' safety in ASCs, heating, ventilation, and air-conditioning systems must comply with all applicable codes when these systems are installed in a facility. Equally as important as installation code compliance is that the facility regularly maintains and services these systems so that they remain compliant for inspections and surveys by accrediting agencies. Facilities must have policies in place defining how and when the systems will be maintained and serviced by qualified technicians, in order to maintain code compliance and ensure patient safety. New heating, ventilation, and air-conditioning systems in ASCs must comply with the 2008 edition of ASHRE 170, "Ventilation of Health Care Facilities," including the design, installation, and performance requirements. This standard also sets out requirements for space ventilation and pressure relationships between adjacent spaces, and parameters for the minimum number of air exchanges per hour, to include the total number and minimum number of outdoor/fresh exchanges per hour. The ASHRE Standard provides specific requirements for those types of space which require 100% exhaust, recirculation of air through units, and minimum acceptable levels of air temperature and air relative humidity.

Temperature and Humidity Monitoring

Facilities should be aware that Centers for Medicaid and Medicare Services (CMS) have specific requirements for monitoring and logging operating room temperatures and humidity levels in order to create an environment that is conducive to good infection control practices. The temperature must be maintained between 68°F and 75°F, and humidity must be at least 20%.

Portable Heating Devices

Because of the risks created by the use of portable heating devices or "space heaters," fire and life safety codes prohibit their use in ASCs. Specifically, such devices may not be used in public areas, patient care areas, staff sleeping areas, or any area that is not attended by staff. There is a long history of fires that have been caused by these devices.

Designated Impairment Coordinator/Fire Watch Policy

Because the fire protection systems installed in an ASC are such an integral part of a building's overall total safety concept, the fire and life safety codes require that facilities must have a contingency plan in place should one of these systems be out of service or not able to function as designed. The systems may be out of service for normal maintenance or because of a failure such as a water main failure or broken component that must be replaced. In order to compensate for the diminished level of protection, the codes require facilities to designate one staff member as an "impairment coordinator," who is responsible both for ensuring that the necessary repairs are made on the deficient system, and for maintaining a vigilant watch of the facility so that both patients and staff will be notified if a fire occurs. Some facilities refer to this as a "fire watch" policy. The codes require this policy to define what the impairment coordinator will do if a fire protection system, either fire sprinkler or fire alarm, is taken out of service. The impairment coordinator must first notify the local fire department of the system that is out of service, and then call a qualified technician to make the necessary repairs in order to return the system to a functioning state. The codes allow the impairment coordinator either to evacuate the building or to conduct a fire watch. During a fire watch, the impairment coordinator must make regular checks for all areas of the building at 15-min intervals, and they must document these checks. Once the system is returned to a normal functioning state, the impairment coordinator must notify staff, notify the local fire department, and can discontinue the facility checks.

Evacuation and Relocation Plans and Fire Drills

Because of the impaired abilities of patients being treated in ASCs, the fire and life safety codes contain very specific requirements for these facilities to maintain policies and procedures for staff actions in the event of a fire or emergency. The requirements include that the facility must have a policy that is in effect and available to all supervisory personnel, written copies of a plan for the protection of all persons in the event of a fire, for their evacuation to areas of refuge, and for their evacuation of the building when necessary. The codes also require all employees to be instructed periodically and kept informed of their duties contained in the plan. A copy of the plan must be readily available at all times in the "telephone operators" location. In addition, when conducting drills, the transmission of the fire alarm signal is required and emergency fire conditions should be simulated. The fire and life safety codes recommend that drills be scheduled on a random basis and that personnel are drilled not less than once in each 3-month period (quarterly).

Required Plan Components
The fire emergency plan needs to contain the following components:

- instructions to first assist any persons immediately threatened by a fire and simultaneously alerting other occupants in the area of the fire by use of a predetermined code phrase;
- instructions to activate the alarm by utilizing the fire alarm manual pull station;
- isolating the fire by closing doors as necessary;
- evacuating the immediate area. Evacuating the smoke compartment, if provided, and the overall facility;
- preparation of the floors or building for evacuation; and
- extinguishing the fire.

Fire Drills
It is important that ASC staff understand that fire drills must be conducted quarterly, utilizing the facility's fire alarm signaling devices. In order to prevent the local fire department from responding, staff must call the monitoring agency that monitors their facility's fire alarm system and inform the agency that the facility will be conducting a fire drill. The monitoring agency will then place the system on "test," allowing the occupants to activate the alarm system without a response from the fire department. The facility staff should also familiarize themselves with their fire alarm system's reset procedures. This information can be found in the owner's manual or by contacting a fire alarm company which can instruct the staff in these procedures. Once drills are completed, it is imperative that these are documented, along with a critiques of the drills, comparing staff actions to the facility's written policies and procedures.

The critique should be used as an instructional and educational opportunity to improve future performance.

ADMINISTRATIVE REQUIREMENTS

Because fire and life safety codes and laws are constantly changing, it is recommended that facilities review their policies and procedures annually.

CHAPTER 13

Staff Fire and Emergency Training Requirements

Because it is so important that ambulatory surgery center (ASC) staff members take appropriate actions during an emergency, it is critical that all staff personnel participate in regular training regarding the facility's policies and procedures to be followed during such an event. A successful outcome during an emergency hinges on the appropriate actions of the facility staff. This is why fire and life safety codes require certain types of training to take place and prescribe the minimum time intervals between training sessions.

FIRE DRILLS

In addition to requiring a written fire emergency plan to be prepared for a facility, fire and life safety codes require that staff personnel hold regular fire drills. This is especially important in an ASC because the patients receiving care at a facility may not have the ability to self-evacuate during a fire. The purpose of fire drills is to ensure that the staff are familiar with the plan and understand their individual roles when a fire is discovered. Staff should be able to implement the plan without direction from supervisory personnel. This includes a basic response to a fire which involves:

- removal of occupants;
- transmission of an appropriate fire alarm signal to warn other building occupants and summon staff;
- confinement of the effects of the fire by closing fire doors to isolate the fire area; and
- relocation of patients as detailed in the facility's fire safety plan.

In an ASC, the codes require fire drills to be conducted quarterly on each shift, with the signals and emergency action required under varied conditions. Fire drills should be scheduled on a random basis while still fulfilling the quarterly requirement.

Ambulatory Surgery Center Safety Guidebook. https://doi.org/10.1016/B978-0-12-849889-7.00013-3

EMERGENCY DRILLS

ASCs must also be prepared for other types of emergency events or disasters, such as tornados, hurricanes, winter storms, earthquakes, acts of terrorism, and active shooter events. Facilities must ensure that they are prepared and have written plans in place to address the types of events that could affect them. Some of these events may require evacuation and others may not. It is important that facilities conduct a risk assessment, and then develop a plan to follow in the event of one of these emergencies. Once a plan is developed, staff must be trained and drills conducted, not only to ensure staff members know the plan but also to ensure that the plan is realistic for the facility. It is not unusual for a plan to be over-ambitious, far beyond the realistic capabilities of staff on hand. Drills will identify what works and what does not work. Often local emergency responders, police, and fire departments will be willing to assist in the preparation of and participation in emergency drills.

PERIODIC REVIEW OF PLANS

It is important that facilities review their fire and emergency plans regularly to determine if these are current, relevant, and appropriate. If they are not, revisions must be made and communicated to staff. Training then needs to take place and drills should be held that will incorporate the changes, so that staff are familiar with the plans and proficient in their execution. It is recommended that a facility's fire and emergency plans are reviewed annually for appropriateness.

ADMINISTRATIVE REQUIREMENTS

The fire and life safety codes applicable to ASCs require training to take place regarding:

- fire drills (quarterly);
- fire drills utilizing the facility fire alarm system (annually); and
- emergency/disaster drills (annually).

The fire and life safety codes require ASCs to maintain documentation regarding training in:

- fire safety plans;
- fire drills; and
- emergency plans.

CHAPTER 14

Surgical Fires

SCOPE OF THE PROBLEM

An estimated 550–650 surgical fires occur every year in the United States. A surgical fire is defined as a fire that occurs in, on, or around a patient who is undergoing a medical or surgical procedure. Despite this recurring problem, there are currently no specific requirements in the fire and life safety codes that address the issue.

The reason fires occur in a surgical setting is that the three basic components required for combustion to occur are often present; however, the surgical staff may be unaware of the risk in this environment. In order for combustion to occur, three elements must come together: an ignition source, oxygen, and a fuel. In a surgical setting, oxygen being administered to the patient can create an oxygen-enriched environment. Fuel is present in the form of draping materials and bedding. These materials may not be thought of as fuel, but when they are basically infused and enriched with oxygen, they provide fuel for an intensely burning fire. The ignition source is often lasers or other cauterizing equipment. When all three elements are present during a medical or surgical procedure, an extremely intense, fast-burning, "flash fire" can occur. The three elements (ignition source, heat, fuel), are referred to as the fire triangle, a concept taught to firefighters in their basic training. To extinguish a fire, one of these elements must be removed. Surgical staff must be aware of this risk and the fire safety concept in order to prevent surgical fires from occurring.

PREVENTION MEASURES

A basic understanding of the elements required for combustion to occur is the first step in preventing surgical fires. Another important step involves conducting a risk assessment at the beginning of each procedure. The highest risk occurs in procedures where the ignition source is used in close proximity to the delivering of supplemental oxygen (e.g., surgery on the head, neck, or upper chest).

Ambulatory Surgery Center Safety Guidebook. https://doi.org/10.1016/B978-0-12-849889-7.00014-5

Staff should also use oxygen safely. They should deliver only the minimum required, consider utilizing a closed oxygen delivery system, and use draping techniques to avoid accumulation or pooling of oxygen in the surgical field. Alcohol-based skin preps also contribute to the potential for a fire to occur. When applying preps, staff should avoid pooling, allow sufficient drying time, remove alcohol-soaked materials from the prep area, and ensure that the skin is dry before draping the patient. Staff should take steps and follow procedures to use all equipment that could be an ignition source safely. The surgical team should communicate the risks present and plan how to manage a surgical fire.

CHAPTER 15

Model Documentation System

MEETING COMPLIANCE REQUIREMENTS

As part of the overall total concept embraced by the codes, there are requirements for documentation that must be completed and retained at facilities. In order to comply with fire and life safety codes, it is imperative that ambulatory surgery facilities maintain all required documentation of inspection, testing, and maintenance of systems, as well as documentation of policies, procedures, and training of staff on site. To be compliant, a facility must adopt a systematic process that will ensure that all required documentation is being completed and at the same time achieving code compliance. This will result in a facility that provides the highest level of safety for both patients and staff.

Importance of Documentation

The importance of documentation of inspection testing and maintenance cannot be overemphasized. As the saying goes, "If it isn't documented, it didn't happen." This is very true when demonstrating fire and life safety code compliance to state or local fire inspectors, or to Centers for Medicaid and Medicare Services (CMS)/accreditation life safety code surveyors. Documentation is required by code for inspections, testing, and maintenance of required building systems and components. This documentation not only proves code compliance, but can assist in creating a historical record for maintenance purposes and can identify trends that may signal future problems with equipment. This can assist a facility in preparing for future maintenance and replacement of equipment and/or systems.

In order to achieve compliance, facilities need to adopt an organized, systematic process for documentation. There are many various logs and checks required; therefore, if the information is not organized systematically, such a vast amount of information can be overwhelming for staff. It is also important to keep the information in a centralized location, whether this is a binder/notebook or file, so that it is readily available for staff to reference and for an inspector or visiting surveyor to access.

Ambulatory Surgery Center Safety Guidebook. https://doi.org/10.1016/B978-0-12-849889-7.00015-7

DOCUMENTATION SYSTEM

Twelve-Month Calendar Framework

One method to ensure compliance is to follow a systematic framework based on a calendar year, with each of the 12 months of the year focusing on a specific area of required building components or systems. This framework ensures that annual checks are performed for that specific system, so that they are not missed or forgotten, while at the same time performing regular required inspection, testing, and maintenance on the other systems and components. For example, April could be designated as "Fire Alarm Focus Month." The facility would complete all other building system and components checks and inspections as required, but the fire alarm system would be the focus. The required annual inspection, testing, and maintenance of the fire alarm system would be completed by a qualified technician. Also, any special work to be completed on the fire alarm system would be scheduled for its "Focus Month"—in this case, April. By following this framework, an ambulatory surgery facility will ensure that annually required inspections, testing, and maintenance will be completed.

In addition to the monthly "Focus," there are ongoing specific weekly and monthly inspections and checks that must be documented for the facility to remain code-compliant. Again, it is imperative that these be completed on a weekly and monthly basis, documented, and retained.

The following is a list of items that should be included in a facility's weekly checklist in order to ensure compliance with fire and life safety codes:

General Fire Safety
- Are area means of exit:
  - clearly and correctly marked?
  - unlocked from inside?
  - clear of debris and equipment?
- Are fire extinguisher inspections up to date?
- Are the extinguishers accessible?
- Are floors and walls free from holes, and clean and dry?
- Are ceiling tiles clean and dry?
- Are fire doors propped open?
- Do all fire doors operate as required?
- Do all trash and linen chutes latch?
- Are combustible materials stored properly?
- Do all personnel know the location of fire alarm pull stations?
- Are there any visible penetrations in the smoke/fire barriers?

Electrical Safety
- Outlet checks:
  - Are emergency outlets properly identified?
  - Do outlets appear in good shape?
- Equipment check:
  - Does all electrically powered patient care equipment utilize the required three-prong plugs?
  - Are power plugs and cords in good visual shape?
  - Do all staff know the procedure to report defective equipment?
  - Are extension cords/multi outlets in use? (List their locations if so.)

Compressed Medical Gases
- Are compressed gases stored only in designated areas?
- Are full and empty cylinders stored separately?
- Are empty cylinders labeled "empty"?
- Are flammable and non-flammable cylinders stored separately?

Generator Weekly Checklist Completed
The following is a month-to-month example, covering all required maintenance, inspection, and testing. This system could be used by a facility to ensure code compliance by concentrating on a particular category or "Focus" for any given month. Remember, the "weekly" and "monthly" checklists must still be completed every week, as well as the monthly generator checks for facilities equipped with a generator, in addition to the focus category items.

MONTHLY FOCUS TOPICS

January Focus—Building Construction
Staff must first ensure that the weekly checks are completed for the entire month. Next, staff must ensure that the general monthly checks are completed. Finally, staff must ensure that the focus items for January, building construction features, are completed as described below:

- Fire walls:
  Confirm all opening, gaps, wall top gaps, joints, are filled with fire caulk.
- Smoke compartment walls:
  Confirm all holes, gaps, wall top gaps, joints, are filled with fire caulk.
- Fire rated windows:
  Confirm glass is labeled indicating fire rating. Confirm glass frame is intact. Confirm glass fits tight in frame and no holes/openings are visible.

- Fire doors:
  Confirm door fire rating label is in place. Confirm door operates properly and latching hardware functions properly. If automatic door closer installed, hardware functions properly and interfaces with the fire and smoke detection system (if applicable). Confirm there are no obstruction impeding operation. Confirm roll down fire doors operate as designed.

February Focus—Means of Egress

Staff must first ensure that the weekly checks are completed for the entire month. Next, staff must ensure that the general monthly checks are completed. Finally, staff must ensure that the focus items for February, means of egress, are completed as described below:

- All required exit doors have exit signs in place.
- Each sign is illuminated and in good condition.
- Where an exit cannot be seen in all directions when standing, exit routes are clearly marked with a sign every 100 ft from the exit.
- Directional signs that appear in exit routes are clearly visible at all times from all directions.
- Passageways and aisles are kept clear and are free of obstructions that could impede movement or create a hazard to exiting occupants.
- The exterior exit path to the public way is not obstructed, and no hazards are present.
- The exterior exit path has functioning lighting.

March Focus—Fire Extinguishers

Staff must ensure that the weekly checks are completed for the entire month. Next, staff must ensure that the general monthly checks are completed. Finally, staff must ensure that the focus items for March, fire extinguishers, are completed as described below:

- Ensure monthly checks of all fire extinguishers have been documented.
- Contact qualified fire extinguisher technician to conduct annual inspection, testing, and maintenance.
- Retain documentation of annual inspection, testing, and maintenance.

April Focus—Fire Alarm

Staff must ensure that the weekly checks are completed for the entire month. Next, staff must ensure that the general monthly checks are completed. Finally, staff must ensure that the focus items for April, fire alarm systems, are completed as described below:

- Ensure monthly/quarterly checks are completed.
- Contact qualified fire alarm technician to complete annual inspection, testing, and maintenance of fire alarm system.
- File annual fire alarm system inspection, testing, and maintenance documentation.

May Focus—Fire Sprinkler

Staff must ensure that the weekly checks are completed for the entire month. Next, staff must ensure that the general monthly checks are completed. Finally, staff must ensure that the focus items for May, fire sprinkler systems, are completed as described below:

- Ensure monthly/quarterly checks are completed.
- Contact qualified fire sprinkler technician to complete annual inspection, testing, and maintenance of fire sprinkler system.
- File annual fire sprinkler system inspection, testing, and maintenance documentation.

June Focus—Emergency Power Systems

Staff must ensure that the weekly checks are completed for the entire month. Next, staff must ensure that the general monthly checks are completed. Finally, staff must ensure that the focus items for June, emergency power systems, are completed as described below:

- Ensure monthly/quarterly checks are completed.
- Contact qualified generator/battery system technician to complete annual inspection, testing, and maintenance of emergency power system.
- File annual emergency power system inspection, testing, and maintenance documentation.

July Focus—Medical Gas System

Staff must ensure that the weekly checks are completed for the entire month. Next, staff must ensure that the general monthly checks are completed. Finally, staff must ensure that the focus items for July, medical gas systems, are completed as described below:

- Contact qualified medical gas system technician to complete annual inspection, testing, and maintenance of fire alarm system.
- File annual medical gas system inspection, testing, and maintenance documentation.

August Focus—Medical Equipment

Staff must ensure that the weekly checks are completed for the entire month. Next, staff must ensure that the general monthly checks are completed. Finally, staff must ensure that the focus items for August, medical equipment, are completed as described below:

- Ensure monthly/quarterly checks are completed.
- Contact qualified medical equipment technician to complete annual inspection, testing, and maintenance of medical equipment.
- File annual medical equipment inspection, testing, and maintenance documentation.

September Focus—Alcohol-Based Hand Rub Dispensers

Staff must ensure that the weekly checks are completed for the entire month. Next, staff must ensure that the general monthly checks are completed. Finally, staff must ensure that the focus items for September, alcohol-based hand rub dispensers, are completed as described below:

Ensure all dispensers meet the following requirements:

- Corridors where dispensers are installed are at least 6 ft wide (clear dimension, exclusive of furniture, supplies, and equipment).
- The maximum individual liquid dispenser capacity is no more than 38.4 fluid ounces (0.32 gal or 1.2 L) in individual rooms, corridors, and areas open to corridors. In suites of rooms, the maximum individual liquid dispenser capacity is no more than 64 fluid ounces (0.53 gal or 2 L).
- The maximum individual aerosol dispenser capacity is no more than 18 oz. (0.51 kg).
- Multiple dispensers are spaced at least 4 ft apart.
- Excluding one liquid or aerosol dispenser per room, not more than 10 gal (37.8 L) of total combined fluid capacity or 1135 oz. (32.2 kg) of total combined aerosol capacity are present in a single smoke compartment outside a storage cabinet. Where a facility uses both types of alcohol-based hand rub dispensers, the sum of the ratios of total fluids/10 gal + total aerosols/1135 oz. should be less than or equal to 1.0.
- Stored alcohol-based hand rub fluid quantities greater than 5 gal (18.9 L) in a single smoke compartment are kept in a container or cabinet manufactured and labeled for such use, or in a wood cabinet built in conformance with NFPA 30:4.3.3(c). Storage cabinets include conspicuous lettering stating "FLAMMABLE: KEEP FIRE AWAY," per NFPA 30, Flammable and Combustible Liquids Code.

- Alcohol-based hand rub dispensers are located at least 1 in. away from any ignition source, horizontally and vertically, including but not limited to the edges of switch and receptacle cover plates.
- If dispensers are installed over a carpeted floor, the smoke compartment involved is fully protected with an approved supervised automatic sprinkler system.
- Alcohol-based hand rub solutions are less than or equal to 95% alcohol.
- The operation of alcohol-based hand rub dispensers complies with all the following (see 8.C.2 for additional alcohol-based hand rub requirements):
- Dispensers do not release their contents without activation by manual or touch-free means.
- Touch-free activation does not occur until an object is within 4 in. of the sensing device.
- The object activating touch-free operation initiates only a single dispensing of solution, regardless of the time the object is kept in the sensing area.
- The dispenser delivers no more solution than required for proper hand sanitation, per the manufacturer's label instructions.
- The dispenser is designed and installed to avoid accidental or malicious activation.
- Each dispenser is tested as per the manufacturer's instructions every time it is refilled.

October Focus—Maintenance Policies and Procedures
Staff must ensure that the weekly checks are completed for the entire month. Next, staff must ensure that the general monthly checks are completed. Finally, staff must ensure that the focus items for October, facility maintenance policies and procedures, are completed as described below:

- Review all policies pertaining to maintenance of the building, fire and life safety systems, and facility equipment to determine relevance and update as necessary.
- Review maintenance contracts and renew/revise as necessary.

November Focus—Fire and Emergency Drills
Staff must ensure that the weekly checks are completed for the entire month. Next, staff must ensure that the general monthly checks are completed. Finally, staff must ensure that the focus items for November, relocation/evacuation fire and emergency drills, are completed as described below:

- Review all emergency and fire drills plans for accuracy and relevancy. Update as necessary.

- Train staff on updated/revised plans.
- Ensure that documentation of quarterly fire/evacuation drills is on site.

December Focus—Staff Training Requirements

Staff must ensure that the weekly checks are completed for the entire month. Next, staff must ensure that the general monthly checks are completed. Finally, staff must ensure that the focus items for December, staff training requirements, are completed as described below:

- Review staff training records and ensure that training has been documented relative to evacuation/fire drills, fire extinguisher training, and updated emergency event policies and procedures.

Printed in the United States
By Bookmasters